Introduction:

This series of essays was inspired by a seemingly simple question. Could flow state state be the result of quantum entanglement across a longer time span than usual? Like many great questions, the process of answering this one led to more, and those to more still. From thinking about quantum theory this way emerged the theory of quantum time, or holistic time. In addition to the initial question of psychological flow, this model offers an explanation consistent with today's leading models for how agency or free will relates to transitions and subjective consciousness, as well as an explanation for the effectiveness of prayer. The three essays to follow focus on each of these mysteries in turn.

1: Quantum Time and Flow State

Time may be the least-understood concept modern physics encounters on a daily basis. But we may possess most of the necessary parts for a better understanding already. This text will attempt to build new meaning out of established facts by considering the implications of four-dimensional holons in this context.

Surprising experimental results and theories exist in fields where time is an unavoidable part of investigations. What has generally been lacking are attempts to construct better models of time and put them into plain language. Given how few scientists in relevant fields write for a popular audience, this reluctance may stem more from the difficulty of building plain-language models than satisfaction with the existing ones. Compounding this lack of interest in relating what happens in the lab to our everyday experience may be the fact that time's behaviour is mostly studied in the field of relativity, while some of the most interesting experimental results come from quantum physics, where time is assumed to be universal and absolute [1].

This text will survey several of the surprising results experiments involving time have produced, and attempt to formulate an intuitively satisfying theory of time based on these results. As such, it will make both scientific assertions and propose how to interpret them. Though the observations and predictions require detailed explanation, the basic assumptions underpinning this model can be summarised as follows:

The first assumption relates to holons. Introduced by

philosopher Arthur Koestler [2], a holon is any level of organization that cannot be usefully understood by understanding its components. It is extremely difficult, for example, to predict the behaviour of a cell by modelling each of the molecules comprising it, even though useful predictions can be made when the behaviour of the cell as a whole is modelled. The cell is a holon: it cannot be modeled as a group of subcomponents without information being lost in the model. Holons of some kind exist at every scale humans have looked for them. Entangled particles may be the most concrete example of holons – their entanglement means that we cannot treat them as discrete particles without losing information. However, while the need for a concept like holons can be proven at such a scale, the concept is more often used intuitively at larger ones. Anywhere we use integers – be it for quarks or galaxies – we are using the holonic model. This model is a prerequisite to the ideas to follow.

Another important assumption that underpins this text is the Qbist interpretation of quantum mechanics. This interpretation describes the observations referred to in quantum mechanics as subjective and only relevant to the observer making them, and that to a third party not observing any part of the measurement process the wave function of a quantum system remains indeterminate until some data from the initial interaction reaches it [25].

Taken together, these concepts of holons and the QBist interpretation of quantum mechanics inform a third interpretation: that everyday, macroscopic phenomena are large scale systems operating on the same rules as quantum systems, and that the same concepts and intuitions are relevant at both scales. This is not necessarily an assertion about behaviour. A baseball thrown across a field and the moon's orbit around the Earth can both be described by Newtonian mechanics, but it would be unreasonable to think we could predict the moon's effect on the tides by observing the behaviour of the baseball. However, different rates of information transfer in a universe described by QBism lends physical validity to the everyday experience of holons. This observation of the rate of information transfer ties into some very interesting theories about the nature of observation and experience such as Tononi's or Penrose & Hameroff's theories on subjective experience – more on this later.

This text therefore asks the reader to consider holons as a serious concept, important to everyday experience in a physical sense, that they stay open to the 'subjective observations' interpretations of quantum mechanics, and that taken together these interpretations allow an intuitively satisfying understanding of how the laws of quantum mechanics express themselves at macroscopic scales. The fourth, and most challenging interpretation this text will present is that holons, which we traditionally think of as three dimensional, also make sense when they are assumed to be four dimensional. This is not a wholly novel idea - Hadley proposed systems of four-dimensional geons as a means of reconciling quantum mechanics and General Relativity [31]. While there is interest in four-dimensional models of quantum mechanics from the physics community as a potential route for reconciling Quantum and Relativistic models, this text is not intended to sort out these cosmic questions at the level of fundamental mathematics. Instead, it aims to present a conceptually satisfying theory of time which borrows concepts from quantum mechanics but is relevant to macroscopic holons like human mental processes. This theory aims to predict otherwise unexpected observations we encounter in both experiment and everyday life, both around time itself and adjacent areas such as causality and perception. Tackling such phenomena necessarily raises interesting questions, which this text will explore without attempting to put to rest. To begin this exploration, we will consider two very strange experimental phenomena that will be useful to keep in mind when thinking with time.

Two particularly confounding experimental results are known as the quantum Zeno effect and the quantum eraser. Though some of the theories and experiments this text draws on are controversial, both of these phenomena are not just highly strange but also widely accepted in the scientific community.

The quantum Zeno effect refers to the fact it is possible to prevent change to a quantum system by repeatedly measuring it in quick succession [3]. We might say this observation confirms that a watched (quantum) pot never boils – certain systems seem to need privacy to change their state and stay static if they are continuously observed. This minimum of privacy necessary for change to take place meets the formal definition of energy, though not of a form we typically think of. In this text we will refer to the energy required to overcome the quantum Zeno effect as transition energy, and while it will be relevant to some of the time-adjacent topics later texts will touch on, it is not immediately obvious to intuitions based on classical physics how not being observed could be a source of energy. One interpretation is that in order to change its state, a quantum system beginning in state 1 must enter into a superposition with state 2 before there is a nonzero chance of observing it in state 2. If the system is repeatedly observed in state 1 at an interval taking less time than it would take to enter into superposition of states 1 and 2, the odds of observing the system in 2 tend towards zero. This minimum time to change state is a holon – its behaviour cannot be predicted by analyzing the behaviour of smaller 'slices' of time, and a quantum system cannot change its state without going through this transition in time unbroken by observation. This image of a system with known initial and final states that requires an uncertain interstitial period for a change in state to occur will be important throughout this article. Note that the times of measurement at either end of the period of change can be known to an arbitrary degree of certainty, but the interstitial period of transition cannot. Unlike a moment of interaction which can be placed at any point in a timeline, this interstitial period of change has a minimum length, and attempting to further subdivide it alters the behaviour of the system. This is what we mean when we refer to a four-dimensional holon.

Another phenomenon that challenges the way we think about time from our classical assumptions is that of the quantum eraser. The quantum eraser is a variation of the classic 'double slit' experiment [4]. In this version, a photon is split into two, lower-energy photons on its way to the standard double-slit wall. One photon continues as in the standard experiment, to pass through the wall and either leave or not leave an interference pattern on the far side. The other must travel further to a reading apparatus and, to the degree it can be said of a photon, spends more time in transit.

The results could be called counterintuitive. Because the photons are entangled, measuring the behaviour of the one that travelled further gives information about the other. This is almost expected – we are used to situations where knowing part of a system lets us know about other parts with which it was entangled. However, whether or not the second photon that travelled further is observed or not directly impacts whether or not the first photon shows an interference pattern, even though the first photon should have already finished its movement by the time the second photon has either been observed or not observed.

A classical sense of time has no way of predicting this result – we do not think of causes as having effects before they happen. If we wish to retain this axiom – and this text will not argue we should discard it – then we cannot think of one event happening 'before' the other. We cannot 'cut up' the chain of events into individual links – both photons having their respective interactions are part of a whole that cannot be further subdivided along spatial or temporal lines. They are a holon that spans not just their respective locations in space but the span of time between the instant they are split and the instant one is measured.

The concepts of quantum erasers and the quantum Zeno effect demonstrate that time is not, as we think of it, infinitely divisible. It is divisible, certainly – interactions can usefully be thought to happen at precise instances – but there are also stretches of time that are more accurately described as being extant. Systems that are extant in time display certain behaviours only if they are allowed to exist on a continuum that stretches across multiple points in time, and we must imagine the time dimension as containing not just points, but fields.

Experiments such as these don't mesh well with most mainstream, intuitively adopted models of time. There are, however, several theories that, while not typically used to examine time, seem to better fit the data. Though this text is focused on better intuitive models of time, it is valuable to study these theories in the more narrow sense their authors presented them in before seeking to relate them to everyday experience.

One of these theories is that for a theory of quantum gravity to give rise to Einstein's theory of special relativity, it is necessary to posit a 'general boundary' to one's models rather than presupposing an infinite spacetime [6]. Key to his approach was the idea that for such a bounded region, as the originator puts it, "Transition amplitudes are associated with regions of space-time and states are associated with their boundaries." Here we see the beginnings of a hard-data model in which definitive, three-dimensional states exist only at the boundaries of four-dimensional regions of spacetime, with the four-dimensional interior being represented by a linear combination of states. Though these boundaries were proposed to ensure that quantum physics display the same 'locality' of results as relativity, the concept is similar to the 'temporal holon' concept this text leans on so heavily.

Penrose and Hameroff's Orchestrated Objective Reduction theory similarly posits bounded spacetime. Their attempt to address the hard problem of consciousness posits that a given superposition has its own spacetime curvature, with each instant of consciousness (AKA experience) happening when more than one of these curvatures interact [7]. Similar to Oeckl's theory, this model posits a collection of bounded regions whose internal states are quantum but whose interactions follow a Newtonian model.

Tononi arrived at a similar model through studying consciousness and neural structures [8]. This model is mathematically rigorous, relating subjective experience inside a given boundary to both the amount of information it generates and the degree to which the information integrates. Tononi defines the first as the scale at which uncertainty is reduced, using the example of a dice which has six possible states while tumbling, which are reduced to one when it lands. This image of uncertainty being reduced relates very closely to descriptions by theorists like Oeckl and Penrose of a system being in superposition across a four-dimentional continuum before 'collapsing' into a three-dimensional state. How Tononi describes the degree of integration of information requires more abstraction. To explain this concept the system in question is modeled as a network of nodes that have values at the nodes and connections between them. The number of different values represents the total data the system can contain, and the degree to which a change to one value changes the values of the nodes to which it is connected represents the degree to which the information is integrated. Specifically, it is the amount of change that the nodes with the fewest connections to all the others have that determines the degree of integration. This description allows the model to determine the proper boundaries of a conscious system, because if nodes of this type are embedded in a system that can pass more information across a less sensitive link, the 'experience' of the larger system is indistinguishable from the subcomponent. Tononi's research focuses on multiple subregions of the brain, but with a bit of speculation it also predicts an embedded model of consciousness, for example the concept of egregores. If a group of people is able to integrate a lot of information between the members, then it meets the definition of being conscious – however, the least sensitive links will typically be between the team members, so it will not be the same kind of consciousness as occurs within the individual members. Within the individual team members' brains, the least sensitive connections will typically still have more 'bandwidth' than the members can exchange between each other through practical communication. If the 'bandwidth' of the communication between members were to exceed that of the members' internal divisions, their consciousness would no longer be functionally separate from one another. Or at least, the separation would no longer be where we would intuitively draw it – if a team's major divisions in communication are e.g. along ideological lines, the cleanest place to subdivide it might be, not at the boundaries of the team members' bodies, but inside the brain of the member most undecided between the ideologies driving this internal division!

We should mention that Tononi's work deals with brain regions at the scale of neural columns and individual neurons – being mathematically rigorous, it is not directly applicable to speculation about the scale presented here. However it is relevant because, by formalising the meaning of individual experience, it can help us think more precisely about what a holon in our context is. A holon in the philosophical sense is a fairly loose concept, able to be conceptualised somewhat arbitrarily. Tononi's work, by associating individual consciousness with the most information passed across the weakest internal link, offers the prospect of an objective boundary for an individual consciousness which corresponds to our intuitions about where a holon's proper boundaries are. This quality of subjective consciousness will be important later in our exploration.

Entanglement is commonly defined as a group of particles where the quantum state of each particle cannot be defined independently. This already brings to mind the definition of a

Holon as an individual unit. Consider Tononi's definition of a complex in light of concepts like entanglement and relaxation time. By his definition, it would be impossible to know the interior state of a complex by observing its behaviour. Because more information is passed internally (even among the narrowest 'choke point' between internal subdivisions) than can be expressed externally, it is impossible to definitively deduce internal state via observing behaviour. This suggests the entire complex is at least partially entangled from the perspective of an outside observer. Of course, an internal means of recording aspects of its state would practically guarantee that they would become available to the environment eventually. And memory is a very useful mechanism. But having a means of suspending such internal measurement would allow the organism to make use of quantum effects in a neural context, as some organisms are already known to do e.g. during photosynthesis [9].

The clash between quantum and classical physics has the potential for so much cognitive dissonance that we tend to find various means and shortcuts for reconciling them, even at the cost of precision in our models. One of these is to naively separate types of systems by scale. In this approach, systems at very low temperatures and scales are modeled with quantum mechanics, while everything larger and warmer is modeled as a classical system [10]. This separation of responsibility works well for making predictions, but the models remain stubbornly irreconcilable and our intuitions about the world are often at odds with themselves.

This investigation suggests that quantum mechanics are present at every level of organization and is a factor of some relevance in any experiment. How well it reconciles our intuitions about the world at small, cold scales and large, warm ones is an indicator of its quality. When two particles are entangled, we might say that they are properly modeled as a single system and that their respective states cannot be described independently of the other. Although the two particles can each be thought of as separate, their behaviour can only be understood when we take them both into account. This concept of entanglement from physics corresponds directly with the philosophical concept of holons. A holon can only be thought of as a single thing, any attempt to reduce it to its components causes a loss of information which will decrease the accuracy of predictions.

Applying the philosophical concept of holons can indicate when it's impossible to directly observe the interiority of a system, and therefore, when quantum dynamics could be 'trickling up' as they do during photosynthesis. Any system which cannot be accurately modeled as the sum of its parts displays behaviour particular to its own level of organization. The inability to observe all components of such a system indicates the impossibility of ruling out the significance of quantum behaviour among those components.

It may be more logical to take a gradated approach to the scaling problem than a sharp segregation of quantum and classical models. Which model is applied to a given system is typically a question of what is most functional, and it is not always necessary to choose between alternatives as though it is a binary choice. For example, in fluid dynamics, whether a flow is turbulent is determined by a nondimensional parameter called the Reynolds number. This number is given by the equation

$$R\_e = \rho \vartheta L / \mu$$

Where R_e is the Reynolds Number, ρ is the fluid's density, ϑ is the velocity of the fluid, L is the length of the airfoil or diameter of the pipe with which it is interacting, and μ is the dynamic viscosity of the fluid. Fluid flow is generally considered laminar if the Reynolds number is below 2300 and transient or turbulent when above. I had always thought of turbulence as an objective measure, as a turbulent flow 'looks' different from a laminar one when observed by a dye or smoke stream. It was quite a paradigm shift for me when, in speaking with a mentor, he casually pointed out that the reference length L is arbitrary, able to refer to an entire wing of a plane or the wing's smallest adjustable flap. A single flow could therefore be both laminar and turbulent, depending entirely on the size of the object with which it is interacting. Turbulence is only a useful concept in relation to a specific solid body with which the flow is interacting, and perturbations that are turbulent to one object could be laminar to another.

Einstein's theory of relativity has demonstrated that the relationship of the observer to what is being observed can influence how an experiment turns out [11]. Despite how widely this theory is accepted, many of us continue to habitually think in terms of a single truth, as though there is an unimpeachable observer who could, if queried, provide a consistent answer to all questions about the state of the universe. Although this premise underpins much of classical physics, we no longer accept it if we wish to form an intuitive understanding of how time really works given the theory of relativity. If there were a single unimpeachable observer, the effects that only take place in the absence of an observer could not happen.

There's no obvious way for an experiment to distinguish whether the collapse of a wave function is an absolute fact caused by observation or whether the distinction between superposition and collapse only has meaning relative to an observer. To think of entanglement as an absolute truth is slightly simpler, but something being true only relative to an observer are not so far outside of our daily experience. Momentum, for example, is commonly understood to only be a meaningful concept given an observer with its own momentum. It may be helpful to take a similarly relativistic view of wave functions. Just as an identical fluid might appear laminar or turbulent depending on the pipe through which it flows, the best model for a system might be quantum or classical depending on the nature of our interaction with it. A more useful approach may be to think of systems not as either entangled or classical, but as a density function similar to how we visualise quantum systems.

Consider a density function representing the possible states of a system: the more precise our observation of it, the more we can treat this density function as though it were a single value – but there is no time at which our measurement error changes from nonzero to zero. Whether uncertainty is relevant to our inquiry is an arbitrary choice. We must first decide how many significant digits we want to make predictions about, and only then can we determine whether the uncertainty in the system is large enough to affect it. When we work at quantum scales, the uncertainty thrown up by the probability distribution is typically larger than our least significant digit. At classical scales, it is typically smaller. But uncertainty is never eliminated as we change scale, there are scales where it can be safely ignored and scales where it can't. Whether a system is quantum or classical is not a function of the system, but of what we are measuring or observing about it.

McFadden and Al-Khalili present a very interesting take on the so-called 'scaling problem' in an article on adaptive mutation [12]. Drawing on Zurek's expression of the time scale over which quantum coherence is lost [13], they presented a possible explanation of how adaptive mutation can cake place, despite the low probability a given adaptive mutation would have. Central to their model is the idea of relaxation time, or how quickly the energy of a particle dissipates due to its interaction with the environment - the time it takes to go from superposition to 'fully observed'. This variable is also important to the quantum Zeno effect.

This concept may be important to the question of how quantum mechanics, which clearly determines the behaviour of the smallest building blocks of our world, affects the world of our everyday experience. A marble rolling off a table and the moon orbiting the earth produce radically different phenomena while obeying the same laws of gravity and inertia. In the same vein, we should expect the laws of quantum mechanics to be as present at our scale as they are at the scale of single atoms, without expecting behaviour to be identical. The variability of relaxation time with scale, and the necessarily partial nature of (de)coherence at larger scales, gives us a means of visualising this transition more smoothly, as opposed to the hard transitions between quantum, classical, and relativistic behaviour we typically assume when considering different scales of physics.

We tend to think of particles as being purely in superposition, and then 'decohering' into a single, classical position following an interaction. However, this idea again assumes there is a single infallible observer who keeps track of the 'true' state of the particle. In fact, each observer has a perspective, and what that observer measures can itself be in superposition from the perspective of another observer [14]. Consider a collection of particles, each with its own perspective. They are isolated and, therefore, in superpositions of various possible states. The higher the temperature of the collective system, the more likely any given pair of particles is to interact, resulting in a shorter time to decoherence. We could also say, the higher the temperature the more quickly each particle's energy dissipates through the system as a whole.

Consider how this would appear from the perspective of the particles, in the case studied by Mcfadded and Al-Kalili a group of protons in a DNA chain. A proton is isolated and in superposition. It interacts with another proton, causing both their wave functions to collapse – or as we might prefer to conceptualise it, their interaction takes place in a particular way, thus limiting the possible future states either can display.

Now consider a third, more distant proton which has not yet interacted with either of the first two. From its perspective, both of the first two protons are entangled and in superposition. When it eventually does interact with one of them, some of the information transmitted to it will relate to the history of these two particles' interactions. From a fourth proton's perspective, all three would have a probability density function that resolves into a particular position and possibly a particular history when they interact. The time required for a particle's energy to 'dissipate' is essentially the time required for it to interact with its environment to such a degree that it leaves a 'signature' in the ambient heat. The 'relaxation time' of a set of coherent particles is a measure of how quickly their energy differences dissipate[12]. That a beneficial mutation would leave more of a signature than a random one has interesting implications for the process of evolution.

Consider this quantity in the context of the scaling function. Energy within a system behaves differently depending on scale. A system comprised of many particles is more likely to interact with a different system, but its exact state will not be determined as quickly. To put it another way, a larger system takes longer to make a unique signature in its environment – it has a smaller 'surface area' relative to its internal complexity. The larger a system is, the easier it is to observe it partially and the harder it is to observe it completely. Much more data can be gathered immediately, but there is also more uncertainty behind what is initially obvious.

If we think of scaling in Quantum Mechanics as a function of energy dissipation, we can intuit how scaling changes the nature of quantum effects. The larger the system, the more likely there is to be a partial interaction, and the less likely there is to be a complete interaction. Some internal entanglement is likely to remain in place, even as some surface level interactions reduce the size of the possible state-space. This approach therefore offers an intuitively satisfying means of scaling up "bounded region" theories of quantum mechanics as discussed by Oeckle, Hameroff, and Penrose.

It is important to remember that this reduction in possible states happens from a perspective. An interaction between a system and an observer reduces the system's possible states to the observer, while a third party that has not interacted with either system or observer perceives a closed system with a different probability distribution, or a different "interiority".

Many scientific models treat the abstract "heat bath" in which a system resides as a sort of universal observer. This can certainly be useful—it helps the math make sense, and it may be an accurate description of the universe in a fundamental sense—but we must remember that as individuals making our own observations, that perspective is not directly available to us. When we speak of "interaction," we typically refer to two holons like an experiment and an investigator interacting with one another. Interactions between a holon and the "heat bath" in which it exists is another matter, The universe may have access to a lot of information about the experiment via the heat bath, but only a small portion of this would be accessible to the investigator. Energy diffuses from a holon to the environment as a function of surface area relative to volume. A large, complex holon such as a human will always be in a state of partial diffusion: some aspects of the holon will be observable to other holons, while other aspects will be purely "internal" over their duration with only their tertiary effects ever available to be perceived by outside observers. Holons are, in some ways, defined by their rate of exchanging information internally compared to the outside world.

Internal interactions still take place in holons of course. In McFadden and Al-Khalili's example, protons in a DNA molecule would still jostle their neighbors, but precisely how this happens is better described as a probability distribution than through classical modelling.

Our classical view of time probably causes much of the trouble we have when trying to think intuitively about matters of quantum physics and scale. We are somewhat accustomed to thinking of holons that extend through space as not quite infinitely divisible. This is an aspect of what it means to be entangled: the subcomponents of an entangled system have some relation to one another, and if they were to be treated as separate and unentangled, relevant data about the system would be lost. In a quite literal sense, the entanglement of a system is synonymous with the system being a holon.

Holons can also extend and be entangled through time, as experiments like the quantum eraser show. Our conventional way of thinking about time as infinitely divisible, much as we used to think about light, is therefore probably incorrect. Systems can no more be infinitely divisible by location in time than by location in space – that is to say, we might specify spans of arbitrarily small divisions, but some quanta will extend beyond these boundaries, and we will lose potentially important information if we try to 'divide up' a holon without taking into account its own boundaries.

Working with the model

This may not seem too significant an adjustment in the abstract; instead of a three dimensional space that can be analyzed arbitrarily within those three dimensions we now work with a four-dimensional space. Calculations of 'surface area' as they relate to energy dissipation and its effects on decoherence, take on a different dynamic, much in the same way modelling heat diffusion on a plane is different from modelling it in a 3D space. But mathematically, a four dimensional space is not terribly different from a three dimensional one.

In the world of physics however, things look rather different. The states of a system at two adjacent points in time are causally linked in different ways than adjacent points in space. If we consider the probability distribution of a system's state, this distribution varies very differently across the time dimension than across the space dimensions.

We are reasonably intuitively receptive to nonlocality in space. That a particle can have a probability distribution describing its location at any given moment – or that a larger object can have a nonlocal aspect to its state which dissipates as more energy is exchanged with its environment – is fairly easy to visualise. Extending this nonlocality to time however breaks down our intuitive image of a probability cloud with various 'densities' corresponding to the likelihood of interactions occurring at that place. If for no other reason, visualization becomes harder when including the time axis because the state at a particular point in time is highly dependent on its state at adjacent points.

Consider the proton in a DNA molecule from our earlier example. With no interactions, its state is a probability density, with any interaction it has leaving an 'imprint' on its local environment. This 'imprint' narrows down its possible states to an increasingly determined, classical state. We also refer to this process as its energy diffusing into the environment.

At the moment of each interaction however, our particle of interest collapses not only its state at that instant but reduces the possible states at adjacent points in time as well. These states are not as widely restricted as the state during an interaction, but we might think of these states as the narrowing of a stream. Any bit of water must pass through the narrowest point – the actual interaction in our metaphor – and the farther away from this narrowing in the flow, the more potential variation there is in the water's position.

It is fairly easy to think of a particle's position as a point within a three dimensional probability cloud of heterogenous density. If we represented a cross-section of these likelihoods numerically, we might see a matrix of probabilities representing the likelihood of encountering the particle in a particular place, like this:

[.05,.1,.05]

[.1,.4,.1]

[.05,.1,.05]

And this causes little trouble to our intuition, as such things go. We are comfortable envisioning a 'haze' of probability around a basically classical universe. But adding a time dimension makes things much more difficult to visualise properly, since this 'haze' suddenly becomes very path-dependent. Suppose a system's state changes over time in an unknown direction, diverging from the normal distribution depicted above:

[.1,.05,.1]

[.2,.1,.2]

[.1,.05,.1]

Now suppose this density distribution cloud were to encounter an outside force which reduced its state. The interstitial probability distribution may 'in fact' be bimodal, but suppose this sort of continuous history was inaccessible to us. The most reasonable assumption would usually be that the progression was never bimodal, but centered on the position where the particle was actually measured.

In most interpretations of quantum mechanics this probability cloud does not represent merely our best guess about the system but the actual interstitial connection between the known beginning and end states. Without an interaction, any number of interstitial probability distributions can be reasonably imagined.

Our typical visualization of these systems involves zero-dimensional particles with three-dimensional probability clouds, interacting in unquantified and infinitely divisible time. If we think instead about chains of events, we must instead visualise one-dimensional progressions whose probability densities can swing widely depending on previous states. Behaviour can also be strongly impacted by what sort of energy diffusion leaves a partial impression on the local heat environment – in any actual chain of events, the progression between events will be partially constrained by the impression left on the heat environment, but multiple (though presumably countable) possibilities would still 'take place' between events. This is what a holon in time is -the undetermined stretch of nonlocality between knowable events.

The concept of a temporal holon is not that difficult to imagine from the perspective of an outside observer. But recall that example of a proton in a DNA chain. A proton could be in a superposition of states that had and had not interacted with a neighboring particle. It is a simple matter to imagine a proton beginning in one position, going through an ambiguous period of transit, and interacting with helicase in a particular position different from the one in which it started.

Once we are comfortable with dynamics like the Quantum Zeno effect, we can accept this period of transition to be necessarily ambiguous – we might think of the energy dissipation of a particle like a proton preventing it from 'having' sufficient energy to change states. As a result, the transition must be unobserved to connect the first and last states.

But how does this period of necessary ambiguity 'look' from the perspective of the proton in transit? Presumably, multiple possibilities would 'appear' open to it, and the standard linear combination of states would apply [15]. A proton is likely a simple enough entity that it would have no 'memory' of taking a particular path through the transition period, aside from its momentum at the moment it interacts with the helicase. The path it took between its initial and final states will be unknowable from the perspective of the helicase.

Suppose now that we replace the proton in our example with a human or similar agent, and 'zoom out' far enough that only the vaguest details, such as initial and final positions in spacetime, are knowable from the perspective of our observer. We would regard the temporal co-ordinates of the initial and final observations as 'fixed'. However, the relaxation time, that is to say the period during which the agent is in motion and interacting only with its heat environment – is not available for event formation as seen from the observer's standpoint [12].

Unlike a proton a humanlike agent would form memories of the transition time. But like all other aspects of superposition extended through time, they will be relatively fluid – eyewitness testimony is weighted quite lightly in court proceedings precisely because it does not really 'nail down' observed events, being rather changeable and context-dependent [16].

When the scale between agents is sufficiently massive, the meaningful effects the smaller agent can have on the larger becomes increasingly rare. This goes hand-in-hand with the smaller agent being better described as a wave than as a particle. However, entanglement in space and entanglement in time work differently as one 'zooms out'. In space, classical mechanics seem to appear naturally through a law of averages as scales get increasingly large. In time, progressions are highly dependent on possible past states. Suppose a past state is not classically knowable. The longer an intervening holon of time lasts, the more different the boundary states can be; the more unpredictable the event progression is, and the more unpredictable the effect they have. Unlike holons in space, which become more predictable as the scale increases, holons in time become more chaotic as they scale up; from completely static at sub-zeno timescales to truly unpredictable at longer scales.

Classical ideas about causality are insufficient to explain the interplay between observations at different times and the transitionary holons which connect them. As the quantum eraser experiment shows, an interaction can restrict the possible states described by a quantum wave function prior to the interaction taking place. However, a quantum wave function evolving in a particular way also affects the set of future interactions which are possible. Agents at different points in time can interact bidirectionally, rather than unidirectionally as we tend to imagine.

This is confusing to try and model as modelling how a complex system will evolve within even a few time steps becomes increasingly difficult with the complexity of the system. To allow for a given interaction to affect the possible past as well as the possible future significantly increases the difficulty of understanding the dynamics.

This is one reason why the idea of holonic temporality can be so powerful – modelling a holon does not require considering actions outside itself, nor the effects of past causes or future attractors. A holon in time has a consistent logic when taken as a whole, it does not require the modelling of interactions at different points in time. For modelling purposes, it occupies only a single continuum in time which cannot be subdivided without a loss of data – past and future are therefore not meaningful concepts inside a holon extended through time, allowing us a middle step in our conceptualization. Between a naïve understanding of interaction of past-future, and the conditionally bidirectional temporality we seem to inhabit lies a means of modelling via simultaneity.

Simultaneity as the term is used here does not necessarily mean events occurring at identical instances in time. Rather, it refers to processes that cannot be subdivided along the time dimension without a loss of data. Two particles can become entangled before being separated, thereafter remaining a single holon despite existing at separated points in space. In the same manner, a process can become entangled and remain a single holon despite existing at different points in time.

While the unobserved freedom needed to overcome the quantum Zeno effect is probably the most literal example of such a holon, a very abstracted example may be a narrative. Although a narrative can be subdivided, meaning is lost if the subcomponents are analysed in isolation. Consider the example of Chekov's gun [17] across a three-act play. The gun is seen in the first act, making a promise to the viewer that must be kept by the end of the third. What is the meaning of the gun in the first act? The answer is very different if we are analyzing it as separate from the third act. Chekov's gun is a holon – if it is analyzed only in the first act it lacks a payoff, analyzed only in the third it lacks foreshadowing. Like entangled particles at different positions, there are logical divisions between the gun's interactions with the narrative, but treating the different appearances as isolated reduces the information we can draw from them.

The obvious question then is what changes when we take into account the information a larger holon brings. One effect of the larger holon in the case of a narrative is the meaning that is drawn. If we sub-divide a play into three acts, the meaning of each act analyzed individually will be very different from the meaning drawn from the play as a whole.

Meaning is an elusive concept that may be purely subjective. One interpretation of this fact would be that it has no impact on the physical world. Consider however, that it is precisely meaning in the context of continuous time that we weigh when making decisions [18]. Adding information, though it tends to make decisions more difficult, also tends to make them better - again, a subjective measure, but an important one.

Another question centers on how accessible the information which requires dealing with holons is during the span of time bounding the holon. It seems intuitive that an agent inside a holon has the option to interact with the other parts of the holon in such a way as to cause a collapse of the wave function - at least, from its own perspective. Returning to the example of a proton in a DNA chain, the proton has no internal memory – thus, while it might have several interactions from its perspective, an outside observer would only observe such an interaction if it changes the behavior of the molecule.

If the agent is a human engaged in some activity, the memory the human builds of the events would be expected to cause a collapse of the probability-space eventually. Humans are constantly orienting themselves in time, and part of the effect this has is to limit the length of time that could be understood as a holon – and hence, to limit the meaning that can be derived from it. This may be one reason why narrative and storytelling is such a fundamental mode of human cognition – it allows us to extend our sense of causality explicitly beyond what is possible implicitly. It lets us draw out relations and meaning by drawing back from a series of events that might seem disconnected and meaningless while they were being experienced.

It would appear that there is an additional means which humans have access to for interacting with a holon extended unusually far through time. Flow state is a quite mysterious process, with significant controversy and much of the available data coming from subjective reports. One of the few pieces of hard data recovered from this phenomenon is that it is characterized by a reduction of activity in the frontal and medial temporal lobes of the brain, aka transient hypofrontality [19].

This is not terribly surprising to anyone recalling an experience of flow. Time dilation – as can occur during a car crash – or compression, which often occurs during artistic endeavors, are both common characteristics of flow states. However, it is counterintuitive when we consider the feats performed in flow states. Why should our ability to orient ourselves in time be suppressed during times of flow? People in flow often make and act on extremely precise predictions, stories abound of star performers 'skating where the puck is going to be' [20] or anticipating their opponents' moves [21]. Why should this become easier with a suppressed ability to orient in time?

With a traditional model in which time is homogeneous and infinitely divisible we would expect it not to. The ability to imagine other times and reason how conditions at one will affect another should be key to making precise predictions and setting ourselves up for future success. Doing any of these things should be harder if we are unable to orient ourselves in time. But with a holonic model of time, the benefits of transient hypofrontality make a certain amount of sense. If our frontal cortex typically works out how different moments relate to one another, it could be responsible for interaction between them – and therefore in a collapse of the wave function which reduces the possible outcomes.

It is possible to affect the 'path' of a photon passing near a detector and thereby changing the set of possible locations it will arrive at. Not observing a photon in such a situation allows the wave function to encompass more possible states, leading to the spooky interference effects we associate with these experiments. Not observing adjacent 'chunks' of time could similarly allow a single wave function to encompass the potentials of both.

Though somewhat radical, this model of time and flow offers explanation for observations that are unexpected in traditional worldviews. In addition to a flow state's paradoxical heightening of predictive ability while deactivating the frontal lobes, it also offers explanation for why flow, despite offering clear advantages, is still a transient phenomenon which no one seems able to maintain indefinitely. Traditional explanations involve the burn rate of neurotransmitters and the nutritional requirements associated with replenishing them. A time-as-holon model offers an alternative or supplemental explanation – if flow requires the brain to model an entire holon, then there would be an upper bound on how complex it could be before the computational limit of the brain is reached.

The tendency for flow to appear with non-judgmental modes of thought is also consistent with the theory of temporal holons. Flow typically happens in high-stakes situations, 'good' and 'bad' outcomes are both possible and easy to contrast. However, a person in flow does not typically recognize any action they take as mistaken until after the flow experience is over. This could be a simple case of not wishing to spare the mental bandwidth for something unrelated to the goal, but instructions for maintaining flow often include instructions to deliberately not think about mistakes one has made during the operation. That would be of questionable utility if the goal is to conserve bandwidth, but it would explain quite a lot if the important thing is to maintain a mental model with no significant internal subdivisions, such as a doer and a critic.

The subjective experience of flow is simultaneously the strongest and the weakest argument for the temporal holonic theory of flow. It could be called the strongest, because the time-holon theory is a familiar description of how flow state feels. However, flow state is characterized by a reduction of activity in those parts of our brain that model time – it's entirely plausible that our subjective experience of flow is misleading, and that any attempt to study the phenomenon by referring to our subjective experience of it is necessarily working off unreliable data.

Even if our subjective memories of flow experiences are not indicative that flow states arise from temporal holonics, it may still be useful to associate the two. This is because whether or not flow state is an example of conditions in which the wave function of a process expands its range of possible states, it is nearly a given that such conditions exist at some scales. It follows that the 'infinitely divisible' model of time is an impediment to how we model the physics of processes, and that a holonic theory of time would be relevant somewhere, even if it has nothing to do with flow states. Our subjective experience of flow states might at least help us think intuitively about temporal holonics where they do occur. Thus we can approach the same central point from opposite directions: Our subjective experience of flow provides an intuitive description for this novel model of time, and the holonic view of time provides a speculative but consistent explanation of some of the more unusual mental states humans experience.

An advantage of taking a mathematical, data-centric view of the world is that it allows us to describe very complex realities in very simple terms. A prime example is that of describing the world or area of interest as a high-dimensional matrix. A typical system of arbitrary complexity could be represented as a 4-D matrix with each particle having 3 spatial co-ordinates at every time co-ordinate. Of course, such a matrix might not be the most complete or compact means of describing a system. Such a space is, fortunately, not limited to four or any rational number of dimensions. But for ease of visualization we can limit ourselves to four dimensions when we begin to think with this cognitive tool, though it can be applied to very different parameters if the investigator wishes.

A concrete example may be useful here. Consider a chess game.

Any given chess game of M number of moves can be represented by an 8X8XM matrix- a 3 dimensional space describing each square on each turn. Depending on how we count duplicate pieces, each of the 64M points in the space that describe one game would have between 15 and 33 possible values.

Suppose we begin with a simple, 100-move game and count fungible pieces (e.g. all white pawns) only once – in this case there would be an upper bound of only 8*8*100*15=96000 possible spaces describing the various 100-move chess games possible. This is a large number, but not impossible to imagine – we can visualize walking among 960 'Stacks' of games, each comprised of 100 vertically-arranged chessboards showing one possible progression from turn 0 to turn 100.

In fact it would be fewer than 96000, because the majority of theoretically possible states of the board would not occur under the rules of chess. The possible states a square can have at any given time step are sharply limited by the states of the squares around it at the previous time step. In other words, while a given square could be in any of 15 states eventually, the need for a valid progression to that state from the set value it held at turn 0 sharply reduces the number of possible games below the upper bound of 96000.

The sort of 4-dimensional space with which we can describe an arbitrary physical system is very similar to this abstract, idealized idea of a complete chess game. Two significant differences exist however: the first is that being four-dimensional, we cannot visualize it using spatial intelligence. This presents difficulty in specific instances, but in an abstract sense it is still an intuitively satisfying way of thinking about the system in question.

The other difference is rather more significant: where the pieces of a chess game have discrete, clearly identifiable locations, particles in a 3D space are properly represented by either rational numbers or confidence intervals. With our current understanding of nonlocality, it may be more accurate to think of the 'position' of a particle as a normal distribution, which can be made narrower by more accurate measurement but never collapsed to a single location of confidence 1. This adds an additional level of difficulty if we are trying to be rigorous, but as an intuition pump the image of our 'chess towers' should still help us understand our 4 dimensional reality.

Withing this context, a holon model of time can be directly accessed. Consider a space describing a chess game. Given the first and last positions with an arbitrary number of moves in between, we can deduce a small number of possible progressions the system may have undergone between state 1 and state 2. Some transitions are more likely than others, as the moves an opponent makes are always partially predictable, and reducing our opponent's options is an important part of any strategy. This is an excellent image to keep in mind while modelling a more abstract idea of agency. In a game of chess there are factors we control and factors we don't, desired outcomes and multiple paths to them. This makes it a useful metaphor in particular because being two-dimensional, we can imagine progression through time via a third dimension. If we wish to apply this sort of analysis to any system where time is important we will need to return to our less easily-visualized 4D space.

A person experiencing flow is essentially taking a

low-probability path through our 4D space – a high probability of a desired outcome removes the challenge necessary for flow state to arise. At macroscopic scales probability - though important - is so complex   and relies on so many nonrandom factors that it cannot be easily imagined as an abstract, universally applicable set of dynamics.

In our hypothetical 4D space however, probability can be directly represented. Consider a plane with a smoothly distributed probability of finding a particle at any given point, like this:

.03125	.0625	.125	.0625	.03125
.0625	.125	.25	.125	.0625
.125	.25	.5	.25	.125
.0625	.125	.25	.125	.0625
.03125	.0625	.125	.0625	.03125

Setting aside for now what happens when an observation in this space is made, this kind of 'probability density' [22] is a standard tool for visualizing particles interacting. If we replace our grid of squares with a grid of cubes, we have a probability distribution in 3D space. Replace the cubes with tesseracts, and we are visualizing probability distributions through time.

Most of the time, the path we take through such a space can be comfortably broad, the same destination being accessible via various sequences of values clustered loosely around one or more 'ideal' paths. When successfully navigating between a given beginning and end state, the number of possible paths, or configurations encompassing the beginning and end states narrows – possible 'paths' become more tightly constricted. What form this narrowing takes can depend to a significant degree on the form taken by the target end state.

Consider now the necessary prerequisites for flow to occur: one of the most universally accepted precursors is intense mental focus in response to a challenging situation. Flow is thought to occur most reliably when a high degree skill is applied to a high degree of challenge.

To approach a mathematical description of this dynamic, we could represent a skill by the ability to navigate a particular sort of four-dimensional landscape – to connect an initial state to a desired end state via continuous intermediate states.

Suppose then that a given skill challenge requires observing a set of intermediate states between an initial and desired condition. We could invert this description to say that it is necessary to avoid observing any intermediate states that rule the desired end state out. If the skill challenge is easy, the number of intermediate states that can be safely observed is relatively large.

As a skill challenge increases, the number of acceptable intermediate states decreases. The response of someone pursuing excellence in the skill in question is to narrow their focus, putting ever more energy into following the narrow path of intermediate states that connects the initial and desired conditions.

In his book 'The Rise of Superman' [23] Steven Kotler catalogues the remarkable rate of advancement in extreme sports over the latter half of the twentieth century, an excellent study of these 'narrow paths' in a practical field. One interesting aspect of this survey is that as soon as a particular record is broken, the odds of it being broken again by someone else in the same discipline appear to increase dramatically.

Rupert Sheldrake's theory of Morphic Resonance [24] would predict this behavior along with similar dynamics in many other fields. This theory posits that events occurring increase the likelihood that similar events will occur again. If we consider this dynamic in the context of our four-dimensional matrix, we might say that when two states appear adjacent at one strata of our four-dimensional matrix, it increases the likelihood that we will observe adjacent states like them at another strata.

QBism is an interpretation of Quantum Mechanics that may offer an elegant mechanism to explain dr. Sheldrake's observations. QBism interprets much of the behavior of quantum systems as subjective, with states such as superposition representing what an agent could reasonably predict about the system in question [25]. If a set of states have been observed in proximity at one point in the past, it stands to reason that if one observes one of them again at a future date, the odds of seeing the others in a familiar configuration go up.

It may help us conceptually to re-examine the nature of nonlocality in this context. We typically think of a particle in superposition as being observable at any number of points in space given a specific point in time. We do not think of its coherence as extending across time, but new effects become conceivable if we do. Entanglement across not only spatial but temporal distances would allow new kinds of information exchange, including the sort of Bayesian predictive models prominent in QBism and Morphic Resonance. To revisit our chess metaphor, if layout L(n) is often followed by layout L(n+1) then observing either of these layouts at a given point along our 'time' axis helps us draw conclusions about the odds of encountering the other, much as observing a particle in space allows us to draw conclusions about a particle with which it is entangled.

From a Bayesian perspective, Sheldrake's theory of morphic resonance would be a natural consequence of extending quantum mechanics through the time dimension without the need for a novel energy field — insofar as QBism is an interpretation that focuses on how we relate to the unknown, consistently appearing next to one another in the past is an excellent indicator that individual states will appear adjacent to one another in the future. Just as atoms are holons that can join together into larger holons like molecules, states are holons that can join together into larger holons like processes.

The more unlikely a given outcome is for a given initial state, the 'narrower' the path between the two becomes. This may paradoxically increase the learning rate for crossing such a state-space, as successfully navigating it once would more strongly adjust the likelihood that any intermediate state which had occurred during the first successful crossing would appear adjacent to the initial state on a subsequent attempt. These temporally adjacent states need not all have been observed explicitly in the past — an observed state can have multiple possible precursor states that could all be in superposition. However, a chain of events leading to an unlikely outcome has fewer possible states comprising it, so knowing an unlikely outcome occurred provides more information about the chain of events leading to it than a likely outcome. This interpretation of QBism, in the context of entanglement between temporally adjacent states, could provide not a mechanism for Sheldrake's 'Morphic Resonance' as well as Kotler's observation that learning happens more quickly when the sequence of events is less likely.

If this version of Temporal Holonics is accurate, it implies a potential method for entering flow. If we view the entire time span containing an experience of flow as a single, coherent holon, then the series of actions needed to successfully navigate the space as desired are one of many possible states that could exist between the initiation of the transition and its conclusion. The skills to potentially take all the correct actions between the beginning and end of a challenge requiring flow need to be present so that the optimal path through probability space exists as potential. To realize that potential requires an ability to enter coherence with past and future 'selves', to make the unlikely 'best' route through the landscape possible, and to observe those states where anchoring the process would be useful. Much as Mcfadden and Al-Kalili posit that useful mutations 'register' in recordkeeping while useless ones do not, lucky breaks and good ideas may 'register' during a flow state while the myriad paths that lead to a bad outcome remain unrealized potential. Careful experimentation with directing attention may reveal a method to reliably trigger such transitions. It would seem likely that if such experimentation bears fruit, much of the technique will involve deliberately not observing the multiple ways an agent could arrive at an undesired outcome.

Humans broadly categorize our own modes of thought along holistic or analytical lines. The current conception of time is particularly analytical in nature, and this 'infinitely divisible' model allows for sophisticated analysis in related fields. But it is likely that such an approach also destroys some of the information that could be present in a more holistic model.

A model in which coherence is extended not just across space but also time, allows a more mixed approach. Viewing some processes as holons occupying bounded regions of space and time opens the door to more holistic approaches to certain problems, and prevents the loss of data that further subdividing them would entail. By recognizing the type of data which is lost if such processes are subdivided anyway, we gain a sense of the kind of additional problem-solving capabilities to be gained by viewing some phenomena as existing across a stretch of time.

One such capability may be an exploratory mechanism for phenomena such as prestimulus response [26]. If we view time not as an infinitely divisible backdrop wherein nothing can affect a previous state, no matter how close they are in fact, then experiments such as Spottiswoode & May's violate our most basic assumptions. If we instead view time as an 'uneven' medium within which some processes cannot be subdivided without losing information, then this class of phenomenon is perfectly in line with our assumptions – some information would exist in superposition across a span of time, and its accessibility 'around' a triggering event is a natural consequence of nonlocality.

If this assumption holds water – and numerous experiments indicate that something of this nature must be in play [27] – then the question arises of whether this dynamic can be usefully employed. Much of the motivation in any study is related to practical applications, and practical applications can demonstrate a controversial theory far more convincingly than any amount of experiment.

Time is often considered external to the experiments in QM [28], and experiments are generally constructed on the assumption that an observation causing a wave function collapse can happen at any arbitrary point in time. Though this is largely borne out by observation, it does not address the fact that observations cannot be made with arbitrary precision, nor the possibility that some processes could resist observation.

The 'localized' nature of time in Relativity is sufficiently well-established at large scales that a theory of quantum mechanics compatible with it would likely have to account for the effect. Phenomena like the quantum Zeno effect are observed at very small scales – but it may be possible to apply the same principles to very large scales as well. Consider the interpretation of the quantum Zeno effect put forth earlier, which suggests that in order to move, a particle has to entangle with a state adjacent in both time and space – and that preventing entanglement by repeatedly observing the particle in one place prevents it from gaining the necessary energy to be observed anywhere but its initial state.

Now consider the creatively named phenomenon of spaghettification. This phenomenon refers to the fact that when an object is accelerated to near light speeds, such as while approaching an event horizon, it appears to 'stretch' along the direction of its velocity. If we apply a similar interpretation to this phenomenon, the logic of the 'entanglement is required for motion' theory would predict that a very fast-moving object is entangled with so many adjacent states along its direction of motion that this 'stretched' set of states appears as a single holon with a 'longer' extent along one direction.

We tend to think of the Copenhagen interpretation as punctuating a period of 'wavelike' behavior with an instant in which collapse reveals an object in a real and perfectly localized state. But in fact, this localization should only occur to the degree that we know the timing and result of the measurement. We tend to think of the 'wave-particle' duality as a strict binary, with a wave pattern 'collapsing' into a local particle upon observation. It is probably more accurate to say that a wave pattern's distribution 'narrows' to fit within the measurement error of the observation causing the collapse. We imagine wave collapse as a normal distribution becoming a 'probability one' distribution where the measurement actually placed it - but it would probably be more accurate to replace the normal distribution of our wave function with a narrower one, bounded by the accuracy of our measurement apparatus. The 'collapse' of the probability wave creates additional certainty, but the new, narrower probability distribution is not qualitatively different from the previous one.

Near-lightspeed objects could be seen similarly as having such wide probability across one dimension that measuring its position along this direction does not collapse its position beyond a certain point – in other words, it 'resists' collapse.

In some sense the question of whether thinking in terms of fields or particles makes better predictions could be seen as an expression of the same dynamic. Rather than see the wave-particle duality as a binary, we might instead say that the nature of the interaction determines whether we see a wider or a narrower distribution of states. A very precise measurement narrows the possible states to a very small set, while a more general measurement allows a wider range of possibilities in the system with which a given observer interacts.

This quality of 'resisting' collapse is important if we are to move from theoretical to practical investigation. Time to decoherence can be calculated given the right combination of conditions and assumptions [29]. If we consider it in the macroscopic sense, it would correspond to the ability to maintain multiple potential states simultaneously. By way of contrast, the ability to enter flow state consistently seems to require not simply existing in a superposition of multiple states across a period of time, but also being able to recognize the preferred chain of events.

The latter seems to primarily require domain knowledge. It's been hypothesized that diligent practice followed by relaxation is a necessary precursor for entering flow state [30]. Anecdotally, flow seems to require visualizing the next step in the process as it unfolds, which is only possible with an accurate domain knowledge.

If a flow state is analogous to a superposition of multiple states and domain knowledge allows us to 'observe' a more optimal 'path' through the possibilities, we would assume something else determines whether enough states are 'included' in the first place. Given that successfully entering flow state requires a large percentage of mental resources to be dedicated to a single task, a similarly large amount of mental energy seems to be dedicated to including potential states in the brain's mental model of the process. If the human brain's functionality lies in partially modeling the universe, it stands to reason that dedicating more mental resources would allow a more complex model – which runs and evaluates more scenarios in parallel – to be made.

Paradoxically, so might the inclusion of fewer time steps. Consider a go player who seemingly wastes a move by placing a stone where it is not helpful to his strategy until very late in the game. Modelling possible progressions of a lengthy game probably requires beginning at a 'fixed' state of the board and diverging the various models from there. Each 'fixed' state so envisioned requires a new model be made starting at the hypothetical state. The more skilled a player is, the farther from a desired outcome he can place his setups, while still accurately modelling the steps in between. Essentially, more skill allows a player to treat a larger block of time as a single holon, with multiple possible states internal to the holon being modeled.

Another codeterminant to flow is creativity, which we might usefully conflate with the introduction of novelty. Human experience is generally that doing anything novel is more difficult than doing something familiar of similar complexity. Sheldrake's model of the universe as forming habits rather than running on static laws [23] would offer an explanation for that behavior. But during flow, introducing novelty becomes easy – or at least, the difficulty thereof is indistinguishable from the difficulty of entering flow.

If we take both Sheldrake's morphic resonance theory and the subjective feeling of easy novelty in flow states at face value, something peculiar must account for their seeming contradiction. An obvious subject for study would be the nature of time in both models. By positing that events tend to repeat once they happen, Sheldrake's theory implies a particular direction to time. Quantum mechanics on the other hand does not posit a direction to time, and experiments like the Quantum Eraser show that this 'bidirectionality' is not purely theoretical.

The 'Holonic' model of time may be compatible with both theories, by presenting time as discrete units which are individually internally bidirectional but may be arranged directionally when multiple ones cluster together. Psychological flow in this theory represents a means of having fewer, larger holons relative to the default mode of cognition, which models the environment as a larger number of smaller and thus less complex holons. By reducing the number of holons in the model, flow allows for fewer, richer, and less predictable interactions between the holons that make up the model.

Suppose we approach Sheldrake's theory of morphic resonance as being limited by the number of steps, rather than a continuous and undifferentiated distance, between an event and points of resonance. With this approach the comparative ease of introducing novelty during a flow experience would be expected – essentially, by extending the active holons along the time dimension, the distance between initial and subsequent instances of a particular pattern decreases.

Such an 'extension' of holons necessarily extends interiority – it ensures fewer interactions with the components comprising a holon at a similar scale. As it is interactions that lead to decoherence, an agent who experiences fewer interactions with the 'interior' of a holon representing some action would expect to experience greater coherence while performing the action. In that sense, entering flow could be thought of as working at a larger 'scale', though the 'size' is only extended along the time dimension. This model is quite consistent with both Oeckl's and Penrose & Hameroff's formulations – and importantly, quite coherent from an intuitive perspective. We might sum up as follows: 'states' only exist at the boundaries of interacting holons. In order for state to change – we might also say, in order for time to pass – regions of space-time must remain bounded. This makes them have no single state inside the boundary, this 'interior' is a linear combination of all its possible histories. To interact with the 'interior' of a temporal holon is to subdivide it; taken to its extreme this prevents any change at all from taking place via the quantum Zeno effect. The larger a 4D holon is, the more potential change is possible between its bounding states, and the more complex its interactions can be.

Though consistent with established theory and intuitively satisfying, there are clearly some unanswered questions that could be asked about this model. For example, what if anything would it predict about group flow?

The boundary of an N-dimensional object has N-1 dimensions. Our default assumption when taking this approach to analyze spacetime is that we take the 'boundary' by omitting the time dimension, this having 3D state-space at either end of this boundary and the 4D transition amplitude in the span between. This is similar to the model already proposed. Suppose now we take the 3D 'boundary' at a space rather than a time dimension and apply it, not to the mathematically approachable world of single particles, but to our human-scaled one.

At the intuitive level, this leads to some interesting images. Suppose our 'boundary' was the behavior of one member of a sports team during a particular play. The 'missing dimension' might be his interiority – in other words, the 3D boundary could be thought of as the 2D surface of his body and/or sporting equipment with time playing the role of the third dimension.

If our hypothetical athlete is in a state of flow, then this internal state is largely left unregistered – fewer or no 'snapshots' of the player's internal state are being recorded in memory or oriented in narrative, consistent with the practice of not judging likelihood of success or failure during the process. Thus the only decoherence happens at the 3D 'surface' of the athlete performing the play. What might enhance or detract from his ability to work effectively with his team?

We might relate the tendency to cause decoherence in a team environment friction. This certainly seems like an effective metaphor for the cost of co-ordination, but insofar as it relates to the rate at which quantum coherence is lost via interaction with the environment, this term may have literal utility as well. If we describe friction as any interaction which raises temperature, then it would stand to reason that this would be the point at which an indeterminable transition amplitude becomes a fixed boundary state. By adding heat to the environment, any source of friction creates a lot of data about the specific interaction of surfaces which caused it, reducing the possible 'internal' states correspondingly.

How then to reduce friction? It seems necessary to specify the time-boundary for this question to have meaning. To continue the sports example, an athlete needs to have a goal in mind which is shared by his teammates – achieving this goal defines the state that the team wishes to encounter at the boundary of their transition amplitude.

Two general approaches to reaching a desired end state with minimal friction suggest themselves. One can reduce the difficulty of each 'step' between the current state and the eventual target, breaking a complex plan into many individual steps that are all simple to execute. Alternatively, one can reduce the number of 'steps' between the current and desired state. Reducing the number of 'steps' or transition amplitudes means that each one is individually more complex and represents more total possibility than a simpler step. With this approach any type of friction, or 'noise' in the data encountered, has a greater chance of reducing the coherence, and thus derailing the plans, of the agent.

Fewer possible paths to a desired end state would naturally mean a lower probability of success, but it would not be a simple matter of more paths to success equaling a correspondingly higher likelihood of success – after all, a 'larger' transition amplitude also comes with more possible routes to failure. What seems a better description is that the more known routes to success exist, the higher the likelihood of successfully 'following' one across the transition amplitude to the desired boundary state.

This begs the question of how a determined athlete or similar subject matter expert can accomplish such successful 'steering'. We may surmise, given the appearance of 'focus' in so many analyses of how to successfully navigate such situations, that the key skill may be better described as the ability to interact with potential sources of measurement so as not to be 'caught' in a state incompatible with the desired end state.

This exposes a possible contradiction in the model – if reducing the possible 'paths' during execution leads to desired results, then maintaining coherence across a larger holon would increase the potential routes to failure at a much faster rate than the potential routes to success. While intuitively sensible – nearer-term goals are evidently easier to achieve – it would seem counterproductive to create 'larger' holons in time for any reason.

One reason our intuition in this case may not square with our experience is the quantum Zeno effect – some minimum transition amplitude seems necessary for any change to take place at all. Extrapolating from the simple example of a particle needing some time unobserved in order to move a minimum distance, we can assume that larger changes between beginning and ending states would require 'larger' transition amplitudes connecting the two. Retroactively the 'history' of the transition amplitude can be inferred – being of a higher dimension than the boundary states, a transition amplitude can contain all the information a series of states would and much more – but during transition, this information cannot always be allowed to propagate into the environment.

From the concept of relaxation time, or the time required for a particle's energy to dissipate into its surroundings, we might say that the greater a potential transition is, the more energy is required for it to take place. Extending the relaxation time of a holon is analogous to saving all this energy for a single purpose, thus allowing more extreme transitions to occur. Achieving difficult transitions of states may therefore be thought of as requiring both a clear idea of the desired end state while reducing the rate of energy dissipation during transition. The less energy is dissipated into the environment, the more impressive the changes that can take place across the holon. This transition energy makes the change possible, but also seems to play some role in 'steering' the movement between the desired states. The nature of this steering will be the focus of a subsequent essay on four dimensional holons and choice.

2: Quantum Time and Agency

This essay will explore how the concept of ego depletion relates to the theory of quantum time laid out in our previous publication. Psychologists as early as Freud and as recently as Baumeister have posited that an individual must expand some energy in order to exercise conscious volition, but the nature of this energy has been only vaguely addressed [1]. Strong evidence for some kind of energy being required in order to exercise conscious volition has been repeatedly demonstrated, but beyond a controversial link to blood glucose levels [2] there have been few attempts to propose a model for this volitional energy.

Ego depletion has not been studied as much as one might expect for such an important concept. One factor in this may be the difficulty involved in properly defining what constitutes an individual in the context of volition and willpower. Nonetheless, the presence of such energy is fairly easy to notice with an ordinary amount of enteroception. Most of us have an experience of using willpower on a task like making a decision or overcoming procrastination, only to find ourselves having more difficulty exercising the same self-control in a similar task later. The expansion of energy from the first task is referred to as ego depletion.

Compare this experience with the 'transition energy' proposed in our previous essay. This energy, associated with a transition amplitude, is associated with the time to decoherence of a quantum system. If there is no energy of this nature present, the system remains 'locked' in place, unable to overcome the quantum Zeno effect. The greater the difference in states at either end of the time dimension of a holon, the greater the hypothesized transition energy needs to be.

This has definite parallels with the energy expanded during ego depletion. The ability to overcome both internal and external conflicting motivations requires an internal coherence – in psychology this coherence is figurative, as we require a 'coherent' mental state without internal dissonance to succeed at acts of willpower. In quantum systems, if we think of a transition as happening when initial and subsequent states entangle with one another, we can speak of literal coherence as being necessary for transition energy to be expanded. As it is expanded, it exchanges information with its environment, much as a quantum system tends towards equilibrium with its environment. Once there is no practical difference between the individual and its environment, there is no energy for the individual to make a change across a temporal holon. There may of course be enough energy of a different kind for changes to propagate further, but they would be predictable to the environment once the holon has exchanged sufficient information with it. A wave pattern is only unpredictable before it is measured, once it collapses it follows predictable rules – there is no transition energy that can be expanded to cause a change that cannot be predicted with arbitrary accuracy.

Although the behavior of large agents like humans are no doubt very different from those of quantum systems like single electrons, there are clear parallels between the randomness of a quantum state in superposition and a human making an autonomous decision. To the degree that a person has free choice in some matter, it is impossible to accurately predict what their choice will be. Executive function seems like a macroscopic version of the transition energy proposed in part 1: it requires an internal coherence and separation from the environment. Overcoming resistance in the environment to enact change involves intense exchange of information, which causes the separation to become less pronounced. Allowing for conflicting patterns in the brain to be modeled as a part of the environment rather than the subject, this would match both the experimental observations and our subjective experience of ego depletion.

It is interesting to speculate to what degree a macroscopic agent's experience of executive function differs from the reality of a particle being in superposition. Although there is certainly a useful metaphor to be drawn, the connection between the two may in fact be quite literal, and based on the same dynamics. Recall that as scale increases, though the likelihood of being observed increases, the likelihood of being observed completely decreases. Traditional versions of quantum mechanics like the Copenhagen interpretation treat the quantum-classical distinction as purely binary [3] while modern interpretations like QBism allow for allow for a more 'partial' measure of collapse [4]. If we allow for partial quantum behavior in our model, we would expect larger complex objects to exhibit it, since surface area increases less quickly than volume in relation to scale. This effect would naturally be more pronounced for four-dimensional processes than three-dimensional objects. And in fact, the field of quantum biology gives some indication that life depends on quantum effects in multiple ways [5].

There is a straightforward way in which the human sense of enteroception and memory could be a factor in how decoherence occurs: if internal states are recorded in memory, they may become available for something in the environment to observe after they have occurred. Mechanisms to adjust this dynamic – of which flow state would presumably be one –

would be expected to have evolved and presumably still be evolving.

There is so much controversy around the capacity for autonomous action that it is probably easier to model what executive function does than what it is. Suppose ego depletion is in fact transition energy being expanded to alter the likelihoods of the associated holon's boundary states. In this case the simplest mechanism underlying this energy expenditure would be the breakdown of separation between the self and the environment. This seems a reasonable first approximation from the viewpoint of a data model like Tononi's as well, since an individual consciousness according to that model is one that can integrate more data internally than across its boundary with the rest of the universe. In this model, communication between the complex and its environment would lead to mutual change, but the more such communication occurs the more data can be integrated across its boundary, eliminating the distinctiveness of its subjective experience and presumably its capacity for individual preference. To sum up, we make the following proposals:

1. Quantum coherence of a system from the perspective of another system is energy in the formal sense, it is a difference across a boundary which can power a change of state, and

2. The executive function of agents at a human scale is an example of this energy, with different dynamics due to difference of scale. Ego depletion is this scale's equivalent of entropy increasing due to wave function collapse (we are here using the thermodynamic meaning of entropy as energy too diffuse to access for a purpose rather than the information theory sense of entropy as uncertainty – the latter would in this

case be decreasing from the perspective of the environment)

This proposal appears at odds with most people's subjective experience unless the model allows separate neural systems to interfere with one another at some times but not others. When a single person has conflicting motivations, the loci of these respective drives are forced to interact whenever the subject of their disagreement is present.

Consider the classic example of a person seeking to overcome a substance addiction. On part of the personality is motivated to avoid the substance in question while another is motivated to seek it out. The subject will be successful in changing their habit so long as the 'avoiding' portion of their psyche has more energy than the 'seeking' portion. When multiple portions of the psyche grapple with the same problem, communication between them naturally increases.

Tononi calls these internal divisions 'complexes' and speculates that their separation is a function of how much information is exchanged between versus within them. As more information is exchanged between versus within them. As more information is exchanged, these division reduce, and less energy is available o make a change – however, this energy typically leads to a new default mode of operation, using aspects of both previously conflicting patterns.

The connection between quantum field effects and the subjective experience of marshalling willpower is most straightforward within the context of a flow experience. Consciousness and the duration of the flow state are both a single holon. The person experiencing the flow state experiences the entire episode as one observation and one decision. Experiencing the entire process as one unit allows working towards the desired outcome for its entire duration. As for what the desired state will be at the outset, the model already begins to get strange here. During a flow state, the subjective experience is typically one of simply 'witnessing' without expanding energy on making decisions at all. However, to enter a flow state at all requires not just high stakes but a high level of domain knowledge. Volition, then, seems to need to be readily available in two forms for flow to occur: On a large scale, in that the outcome must be of obvious import, and a small scale, in that domain knowledge must be sufficient that no individual decision in service to the overarching goal is particularly difficult.

Penrose and Hameroff's as well as Tononi's models of consciousness posit an interiority to bounded regions, with subjective experience an intrinsic quality of this interiority and objective facts an intrinsic feature of their shared borders. Temporal holonics begins with these presuppositions and specifically considers them in four dimensions. This implies the scale of the experience – and therefore the consciousness experiencing it – varies not just with the three-dimensional extent of the substrate's boundary, but its bounded region across three spatial and one temporal dimension. Quantum mechanics, with its existing concepts of superposition, entanglement, and decoherence, provides a convenient language for modelling bounded regions of spacetime. A key feature of this model is that while the states at the boundary of such a region can be known, the internal particulars are not only unknowable, but exhibit very different behavior than would be predicted from observing only sequences of boundary states.

The fact that subjective consciousness is associated with the interiority of holons of this nature further complicates the final model. There is evidence of consciousness having measurable effects on random events, with larger quantities of consciousness having greater effects [6] [7]. As these studies relate to consciousness having effects outside the brains of the subjects, they could be described as investigating holons which include humans. By Tononi's model, we might say these studies relate to holons where the greatest bottleneck to information integration is between humans and the rest of the holon. These studies are relevant for two reasons: they indicate a possible mechanism for consciousness to effect change as our subjective experience using willpower implies it does, and they indicate that 'more' consciousness, whether coherent across space or time, has more extreme effects.

This may have some relevance to how willpower relates to the temporal holonics theory. It has been suggested that humans are unusual among animals because of our felicity at taking not just our present but our future wellbeing into account when making decisions – that is to say, our capacity for delayed gratification. The ability to use memory to inform behavior for immediate wellbeing exists elsewhere in the animal kingdom. The ability to balance immediate against future wellbeing however, is only partially present even in humans. Indeed, much of the research around ego depletion engages with the difficulty humans have balancing short-term against long-term gain. Under the theory proposed here, the degree to which an individual can prioritize long-term benefits corresponds to how long of a time frame they are able to maintain coherence over. One prediction this theory offers is that familiarity with flow states would correspond to a greater ability to exercise self-control in a 'marshmallow test' [8] scenario, since both entering flow states and prioritizing long-term benefits over short-term would relate to extending coherence across the temporal dimension.

Our previous essay presented the hypothesis that a holon can extend through time as well as space, that flow states are an example of holons extended through time, and the unusual results flow sates allow are the result of larger holons having more potential boundary states.

Tononi's and Radin's research both point to an interesting connection in the context of this model. Tononi, in seeking to solve the hard problem of consciousness, begins with a mathematical model of neural networks and finds that a region's ability to integrate more information internally than it can communicate corresponds to its having subjective experience, with the amount of information integrated corresponding to the reliability of consciousness being present. Radin, seeking to explore certain strange results at the intersection of psychology and physics, began at the opposite end – positing subjective attention as a given, he found that certain phenomena sometimes called 'spooky action at a distance' intensify with the amount of subjective attention present, both via number of observers and intensity of focus [9][10].

Taken together, these results point to a conclusion that seems obvious from our everyday experience: subjective consciousness can affect the universe, and the 4D hypervolume of the holon having the experience corresponds to the impact it can have. The method to move effectively between existing and desired end states may be as simple as maximizing the extent of the subjective consciousness which prefers the given end state. There are multiple means to achieve this:

-maximizing the amount of neural subregions which have the

preference and attention.

-maximizing the time during which the subjective experience is coherent

Both these strategies can lead to a 'larger' holon and therefore 'more' willpower.

Various means of achieving these ends have already been widely promulgated in pop-psychology. Some of the more obvious include visualizing the desired outcome (which we would expect to 'spread' the desire across more neural subregions), associating with people who share the same goals (which would add additional ego involvement), training attention span (which would increase the length of time a particular goal can be focused on), or various means of gaining self-knowledge (which would allow subregions of the individual to communicate and therefore 'share' subregions across multiple goals more efficiently).

Another prediction temporal Holonics can offer here is that increasing time to decoherence would significantly increase the effects of the coherent mind having the experience. Recall from our paper on flow that time to decoherence is not an absolute quantity, with some information about a larger subject becoming more easily available, and some remaining obscure for longer.

Focus, a key component of flow, could well be one means of extending relaxation time. The portions of the brain that model an environment and how to complete a task, by engaging the executive function, are required to ignore a lot of input [11]. This includes input from other neural regions as well as sensory input not relevant to the task at hand. Deliberate avoidance of interactions that do not serve the subject's narrowed focus would be consistent with a model of execution where the energy to move between states with few potential connections requires maintaining some degree of privacy.

Applied willpower requires several of these features, involving a narrowing of the worldview and either sublimating noncompeting goals or dropping competing ones. The basic model presented here is of instances of psychological flow allowing more degrees of freedom because their content are a single temporal holon with far more action potential than the sum of several smaller holons. The neural tendency to stop orienting in time during these experiences while still enjoying high predictive power speaks to this model's robustness. Research like Tononi's and Radin's links these 'large holon' experiences to subjective consciousness directed in a consistent way can affect the outcome of quantum processes.

Up to now, this model has only been used to analyze individual, temporally localized holons. There is much interest in how an individual can most effectively pursue a long-term goal since such goals require far more time than it is possible to maintain a single flow experience over. This model is not directly applicable to such questions. However, it may provide insight into the building blocks that would make up an effective method.

In order to analyze such dynamics, we must take a closer look at the idea of nested holons. This idea is an integral part of the concept of a holon. In a classic example, proteins are holons inside cells which are holons inside organs which are holons inside humans which are holons inside social groups... it is obvious from this example that holons are everywhere, and that almost all can easily be subdivided or integrated into smaller or larger holons.

What is missing from this traditional analysis is that holons are not confined to three dimensions - in fact if Oeckl's 'General Boundary' [12] interpretation of quantum mechanics is correct then holons may in fact be not 'just' metaphysical constructs but physical realities, albeit primarily four-dimensional holons with three-dimensional surfaces, unlike the three dimensional holons with two dimensional surfaces we are accustomed to thinking with.

It is in this four-dimensional realm that all human competence takes place, even though human experience seems limited to three dimensions [12][13]. Despite this limitation, it should be possible to conceptualize the nature of embedded actions as they relate to human experience.

To be successful at a complex task, specialization in sub-tasks is often helpful - this allows greater flexibility when different challenges share some aspects. However, integration of information relates to intensity of experience. If a longer task can be subdivided into subgoals and the neural regions specialized for these are fairly independent - that is, if variation in the state of one doesn't lead to much variation in the others - then the intensity of consciousness pursuing the meta goal will be reduced and consequently, so will the intensity of its willpower.

A delicate weave must therefore be achieved: sufficient focus must be brought to each subgoal that the individual 'steps' on the path to victory are arrived at & recognized as such. However, they must also be integrated inside a greater temporal holon. Within this holon they must be capable of exchanging information with the rest of the potential states comprising it, not only offering context to the rest of the holon but changing depending on the context they exist inside of.

Properly engaging in this weave depends on knowing the appropriate relaxation time of the subgoals where the benefits of the movement towards the goal being a holon - transition energy being free to use across a larger probability space - are not disrupted by too many factors from outside the probability space of the outcome. This is exemplified by the concept of 'monotasking', to which many prolific people ascribe large amounts of their success: Focusing all one's attention on a single task until it is finished (or some predetermined quantity is finished) before engaging with another. This intensity of focus allows a great deal of probability to coexist while the person in question engages with the subject. However, if the work is embedded within a wider context - doing a job well being embedded in a career for instance - then some interaction between the various subtasks in service of an overarching goal is to be expected. The neural architecture devoted to the larger goal, when extended through time, is necessarily more diffuse than that of the sub goals. However, it is necessarily also larger. This dynamic presumably exists at every level of organization. How much information to exchange is therefore a crucial question about architecture pursuing a sub-goal.

'Slow is smooth and smooth is fast' is a popular piece of folk wisdom. Within the context of a question about ideal holon-span, it is rather telling. "Smoothness" is clearly a function of individual tasks being integrated into the whole, implying that a longer stretch of temporal holon is preferable, allowing fewer internal interruptions. Similar principles around minimizing handoffs and interruptions are emphasized in lean six sigma methodology [14]. A similar dynamic is presumably at play between different neural regions, with a single, well - integrated architecture being more effective even than more optimized subregions that are not as well-integrated. A core skill the temporal holonics model predicts would be useful is the ability to seamlessly integrate nuanced information across 'larger' holons.

However, some processes are too large to do this in a straightforward manner. Certain projects may benefit from larger individual 'chunks' without being possible to complete in a single session.

When distances are spatial, we take for granted that there may be 'gaps' between parts of a holon. Entangled particles, separated by a significant distance, are obviously still a part of the same holon. The Copenhagen interpretation tells us that when entangled particles are separated in space, observing one 'collapses' the wave-function of the other 'instantaneously'. However, experiments like the quantum eraser have demonstrated that a particle's wave form collapses not 'instantaneously', but more accurately 'retroactively', along the entirety of its transition, not merely the 'edge' where an observation takes place.

Given these distinctions, can we conceive of a temporal holon that is not concentrated in a single temporal locality, much as entangled and separated particles are setpoints of a holon not concentrated in a single physical locality? would a long creative project, to which the creator returns periodically, interspersed by periods of rest and other duties qualify?

By the model presented here, the answer to that question depends a great deal on the interactions between these individual work sessions and the periods in between. The temporal holonics model deals with superposition not as an absolute but a relative quality, with larger holons having less extreme but longer-lasting quantum effects. The rate at which the energy potential of superposition states dissipates into its environment is variable, depending on more factors the larger a holon is.

One phenomenon this model would explain is the reluctance of many artists to show unfinished work to the public. Rather than mere perfectionism and embarrassment at showing off incomplete work, this behavior could be a defense against premature wave-function collapse. Limiting interactions between the incomplete work and the rest of the universe would allow the probability-space bordering the work to expand, thus increasing the quality of the best work that could possibly be produced.

Within the context of an individual work-session, two questions arise: What the ideal length of holon to 'aim' for is, and precisely how one 'aims' for the 'best' of the observable 3D possibilities 'touched' by the 4D holon describing the movement between an initial and one of many possible subsequent states.

On the latter question, a model such as Tononi's would posit that the holon itself is consciousness experiencing the transition, more concentrated consciousness as well as the relative probability of a given outcome should both be relevant to the likelihood of a given desired outcome being achieved.

Bringing more consciousness to bear on a given problem is the obvious way to achieve a desired result. According to Tononi's model this would have two components: That a large neural region is dedicated to navigating the probability-space, and that a lot of information is exchanged internally to this neural region.

Extending the coherence of the neural region through time increases the chances for both to happen. There is one wrinkle however - the region cannot be so large that there is internal disagreement over the 'best' outcome to the transition; this would necessitate internal observation and therefore make it impossible to maintain a coherent holon across a long timespan. All such internal disagreements should therefore be addressed before the subject can expect to do their best work, which may explain why periods of high performance seem to require periods of struggle combined with rest and relaxation [15].

A second practical question arises around the ideal time to spend in a single holon. Some studies have indicated that flow is most likely to occur when a subject is required to perform at 104% of a previous best [15]. A useful experiment would therefore be to determine the maximum time a subject can reliably spend "navigating" a single holon and attempt to exceed that time by 4%.

One difficulty of designing such an experiment would lie with the fact that "sustaining" a single holon is not a binary measurement in the model presented here. Coherence on the scale of a neural region will likely always be partial from the point of view of a human investigator. Determining whether a particular holon is coherent enough to result in a flow experience is therefore the first challenge to solve.

A subsequent challenge is that the two components of bringing more consciousness to bear on a single challenge have somewhat competing prerequisites. The larger a neural region is, the more potential consciousness is associated with it. However, a larger region is more likely to have internal regions across which little information can be integrated - the 'intensity' of a larger neural region's consciousness would therefore be expected to be lower. Prioritizing the amount or intensity of consciousness therefore seems related to the brain's 'Diffuse' and 'Focused' modes [16].

This may be a consistent difference, not just between focused or diffuse neural regions in a single subject, but neural regions of larger versus smaller subjects. Smaller organisms consistently detect and react to stimuli across shorter spans of time than larger animals [17]. This finding is consistent with the model of an agent as the consciousness within a 4D holon whose decoherence time is determined in large part by the size of the agent. A smaller agent would take less time to exchange the majority of internal information with the environment. It would also take less time to process this information internally, having a relatively smaller probability space to explore. A smaller number of internal states for the agent's consciousness to 'parse' means that though it can only rarely navigate to an unlikely outcome, it is able to move to more 'nearby' potions of probability space more quickly than a larger agent.

Humans have an advantage in this regard because we have the perspective to choose whether to focus on rapid readjustments across more, smaller holons or would prefer to aim for the lower-probability edge cases that are only accessible via fewer, more extensive temporal holons. Both are important to high performance - in computer-game parlance these complementary areas of competence are often referred to as the 'micro' and 'macro' aspects of problem solving. Of course, choosing the time span across which to optimize this adjustment is a challenge of its own requiring the same kind of judgement to be made.

the various timespans across which we can orient decision making touches on a challenging concept within this work: The integration between larger and smaller holons, all with their respective amounts and intensities of consciousness.

Tononi's model posits that consciousness within a complex is present when the information it contains is integrated, which he defines as a state where a change of one datum changes at least some of the others as well. Both when considering internal holons and holons to which we are internal, this definition allows us to formulate relevant questions. Although we might initially suppose that more integration leads to more available consciousness and hence more willpower, there are also things we would not wish to be easily altered. Much of the data our neural networks contain, such as our deeply-held principles, are most effective when they are not affected by changes to the rest of our neural structure - when they are not part of the same consciousness, to go by Tononi's model.

Constraining a holon in this manner allows the subject to limit the transition energy of the holon to relevant potential changes, and by narrowing the focus in this way the consciousness becomes more intense at the cost of being smaller.

In their book on willpower, Baumeister and Tierney relate the story of Henry Morton Stanley and his search for the lost Dr. Livingston. Stanley was known in his day for his force of will, and the authors relate the anecdote that upon waking he always shaved before anything else, requiring one fewer daily decision. Consider how this anecdote would be modeled as a group of nested 4D holons: A long jungle expedition is the largest holon, requiring multiple decisions, and having many possible outcomes, though the desired outcome of locating dr Livingston is known from the start and invariant. The question of what to do first in the morning has the potential to affect all other portions of the encompassing holon. Does this necessarily mean that the best strategy is to vary the first thing one does in the morning so as to maximize the potential outcomes? Stanley clearly did not believe so, and Steve Jobs's habit of never choosing a different outfit in the morning seems to echo this pattern. In some sense we might see these habits of reducing the decision space as a means of 'shrinking' the encompassing holon.

In a 'transition energy' model of holistic temporality, the energy to change state relies on the holon being in a state of partial superposition and having sufficiently little interaction with the wider universe that by the time the next such interaction occurs, the holon's probability distribution includes a particular observable. To the degree the holon has agency, we would say it is 'choosing' the state of this particular observable, while the portions not 'chosen' by the holon itself hare presumably amenable to Newtonian analysis.

Within this context, making certain decisions only once and repeating their consequences reduces the probability space the holon encompasses. Steve Jobs wearing nothing but black turtlenecks removed any presentation he might have given wearing something else - an order of magnitude reduction at least. If this approach was not merely ritualistic, perhaps based on a familiarity with Stanley's philosophy, we would imagine this practice would 'constrain' the transition energy Jobs had access to, concentrating it in a way that allows more unlikely changes to occur elsewhere. We might say such reductions in probability space are a means of increasing the power available without altering the total energy.

On more speculative ground, it seems worth investigating whether the uniformity of outward appearances habits like Jobs's turtlenecks or Stanley's daily shave resulted in lengthening the time to decoherence both men experienced in some way. The human instinct for privacy, seemingly at odds with a social environment that frequently rewards visibility, might be an outgrowth of this phenomenon, a desire to concentrate transition energy for some time. Jobs's habit of wearing identical outfits brings to mind a habit some public figures adopt, of wearing identical outfits on different days so as to obscure when a given photograph of them was taken. such deliberate reduction of incoming attention when attracting it is the source of one's income is surprising behavior. But if some amount of transition energy is necessary to properly manage one's affairs, it seems much more reasonable. Reducing the variance between observables at different points in time would create a stronger separation between the information a subject exchanges internally vs externally.

An additional question arises regarding quality of attention: does the 'type' of attention paid affect the quality of transition energy the subject can expand? As noted earlier in this text, in regards to the recursive operation of a human neural net the answer seems to be yes. Certain types of self-observation, particularly those involving judgement of right or wrong, appear to interfere with attempts to experience psychological flow. Others, particularly observation that does not require ordering events in time, seem to strengthen the experience of flow. This observation is entirely consistent with a model of cognition as multiple four-dimensional holons whose boundaries are defined by fidelity of information transfer as per Tononi, with consciousness as humans experience it associated with the volumes and experiences associated with the boundaries. The question is how such a model could be expanded if more than one neural network were in play.

Under the penrose-Hameroff model, as well as QBism, observations are particular to the holons making them and, though typically symmetrical across the two holons whose surfaces interact in the process of observation, can be in superposition from the perspective of holons not involved in the initial interaction. In theory therefore, observation of a process involving, transition energy could enhance the 'size' of the holon, adding more consciousness to the process of navigating probability space, though at the price of lessening the intensity of the consciousness. This would explain why 'clear shared goals' and 'high bandwidth communication' are often mentioned as common triggers of flow [15]. Shared goals would prevent the respective volitions from cancelling one another out, while high-bandwidth communication would increase the intensity of the total consciousness experiencing the holon in question.

The technique of observing team members so as to increase rather than decrease their transition energy can presumably be practiced. Tononi's model can offer insight here consistent with a properly scaled-up quantum system. Recall that per Tononi, what characterizes a conscious system is a higher rate of information transfer within the system than across its boundaries. Following this logic, we would expect that observing a follow agent would tend to increase the consciousness available the more changes our internal state can undergo depending on our observation of them. This of course requires a strong shared trust, since undergoing internal changes in response to another person is psychologically risky. It also requires a sophisticated model of the person as a simple one would produce fewer distinct internal states. This is consistent with another of Kotler's precursor to group flow, deep familiarity. Being willing to update our model, particularly so as to include additional nuance, would further strengthen internal avenues of communication relative to the external.

A key point should be highlighted here: Tononi's model associates a complex with more internal than external information transfer with the phenomenon of subjective experience. Our model posits that given a subject with more internal than external information transfer and an interpretation of quantum mechanics that has multiple relative truths rather than one objective one, such a subject would be in a state of superposition from the perspective of any observer outside its boundaries, and that this transition energy would be necessary to change state. In either model, what indicates this state is active is the presence of subjective experience. Any experiment or skill developed from this model could therefore determine the state of attention by whether the attention paid is objectifying and determining concrete things about the object of attention, or whether it stays subjective and experiences without determining facts or labels.

Observing without conceptualizing is a practise heavily emphasized in some traditional schools of mental training [19]. That such awareness makes a larger holon with an attendant increase in the amount of consciousness experiencing the open possibilities of being in superposition, would make for a plausible modern-day explanation of why this mental state is so heavily emphasized in traditional regimes of mental training.

The relationship between meaning and subjective experience highlights an important distinction. Meaning is subjective - data flowing into a holon may be objective, but the meaning assigned to this data is part of the receiving holon's internal state, associated with its transition energy and subjective experience. In effectively maximizing transition energy shared between multiple agents, shared meaning is therefore vital. Being both highly abstract and central to motivation, divergent beliefs about meaning can severely limit the amount of data being shared internal to a team that affects the state of the rest of the team, as a lot of data can combine to make a relatively small amount of meaning.

When considering what makes for the maximum amount of information exchanged between members of a team, an argument can be made that assuming more than one source of meaning or key motivation in one's members would lead to more potential states than only one. However, it is particularly nuanced communication that gives rise to rich information transfer and a difference in the subjective experience of a holon's hypervolume, as opposed to the objective experience associated with the hypersurface. Conflicting beliefs about meaning render much of this nuance irrelevant. A shared sense of meaning would also imply that the same transition energy is concentrated among fewer alternatives. This concentration of available energy would likely increase the chances the subjects would have of encountering the desired outcome from among the smaller group of likely outcomes. This prediction is similarly in line with Kottler's observation that 'shared clear goals' are necessary for the emergence of group flow [15].

We would also expect to see that subjective observation of one's team members would be present at higher levels of performance than objective observation. We might characterize this difference as emphasizing perceiving one's team members without putting much energy into judging them. At the same time, maintaining the borders of one's team would likely be aided by sharply judging the things the team interacts with; we would expect this to focus the team's transition energy. This implies that the image of a well-communicating team would be one where the team members observe each other subjectively and whatever lies outside their team-likely the world apart from their most important tools - objectively. This resembles the internal state of a properly functioning mind under this model, in which internal communication has the nuance of subjective experience while information coming into it is collapsed into objectivity.

One additional complication may be mentioned: Recall that a holon under this model is four-dimensional. This means that a well-functioning team, able to harness its collective transition energy for difficult tasks, might also engage in objective judgement of its own performance after a period of performance. Athletes often perform in a state of judgement-free flow, only to very critically examine their performance after the competition is over. We would expect a high-functioning team to similarly emphasize periods of subjective flow and performance, punctuated by periods of objective assessment and re-orientation to a new challenge.

3: Quantum Time and Prayer

Building on the distinction between subjective and objective styles of observation described in the essay on quantum time and agency, a key question is how these dynamics alter when holons exchanging information are of very different scales. Consider the interplay between multiple team members seeking to accomplish a complex task. At the smallest level of organization, we have individual team members for the duration of their individual flow experiences. These units of measure may themselves be ambiguous, as a typical flow experience can be more or less intense. It is certainly true that a human experiencing flow does have some interaction with the environment and thus, is only ever partially in a state of superposition. The degree to which an individual flow experience is 'broken up' by effects like being observed or forming memories itself creates a 'sliding scale' of larger and smaller as well as more or less intense flow experiences for the person in question. The individual 'starts and stops' of consciousness that characterize everyday experience are presumably identical in quality. It is only in scale that the holons of the everyday differ from the holons of flow, with probabilities changing nonlinearly as the holons increase in scale.

If we extend our investigation beyond the narrow focus of a sporting event - say to a group of co-workers pursuing a flow-inducing goal over a period of months - then these 'local' holons, personal and group both, are embedded within one or more larger holons. To what degree the same kind of consciousness which characterizes holons in a neural net over a short time period can extend to such larger campaigns, with their starts and stops, is an intriguing question. We may presume that this dynamic follows the same tradeoffs we see at local scales, with a lower 'density' of consciousness commensurate with its greater 'size'.

Three or four distinct scales are thus relevant in considering the dynamics of human consciousness as four dimensional holons. At the intrahuman scale, individual complexes within the human neural network exchange information across short time spans. These smaller units of consciousness are likely more intense but less meaningful than the large narratives composed mostly of memory and prediction that make up everyday experience. Extending a smaller number of possible states through a longer time characterizes flow experience, with access to less likely outcomes becoming possible. Being of a higher dimension, the 4D mind increases in size faster than its 3D 'surface', or experiences.

At the extrahuman scale are holons composed of multiple actors operating over a long time frame - we might call such holons 'campaigns'. The amount of information exchanged internally to such holons is sparse enough that it would be very diffuse compared to a human's consciousness. Nonetheless, so long as substantially more information is exchanged internally as opposed to with the environment, the campaign's internal state would be in superposition from the perspective of the environment and thus the outcomes of its transitions would be unpredictable. The result of this transition energy being expanded would make the campaign its own agent, albeit one with very different subjective experiences than the human scale ones we think of by default.

As for these human-scale consciousnesses, this is where we can count a third and possibly a fourth scale. An individual agent often experiences group flow while being part of a team of people joined for a common purpose. In this case, the individual, more intense consciousness of the person is a 'building block' of a larger, more diffuse consciousness. But the individual subunits of a human agent's experiences typically contain models of the other team members - more accurate models allowing better group flow. Statistically, these other members will consist of temporal holons for only a fraction of the time the largest holon characterizing a team member's experience.

Thus the temporal holons characterizing human experience are reciprocally embedded. A humanlike consciousness can be embedded within a team that is physically larger than itself while modelling and observing the team in a holon with a longer relaxation time than the team itself.

This may be why the image of the fates at their loom is so compelling when we consider the nature of time. Mutual subjective observation extends and interweaves the subjects, while objective observation shortens and distances them. This interweaving based on the kind of attention passing between subjects is a key feature of the interplay of different consciousnesses.

This difference between subjectifying or objectifying - or being subjectified versus being objectified - can be intuited at the human scale. But we would expect that some of the most interesting dynamics would happen when the interaction happens between holons at vastly different scales. Consider the subjectifying-objectifying dynamic applied between the scale of a human and a few human cells, or a human and something comparatively larger than a human, perhaps 'the arc of history' or 'nature'.

Subjective interaction with holons much larger than a human is conceptually very similar to the traditional concept of prayer. If interaction without the creation of facts (IE subjective observation) interweaves the holons in question, then subjective observation of a holon much larger than the observer would facilitate an expansion of the smaller observer's consciousness - and perhaps, also a local intensification of the larger holon's consciousness.

We may consider this intensification of consciousness as bringing in more transition energy or ego that may be expanded during ego depletion. Holding an image of something desired in mind while opening oneself to a much greater holon would presumably influence which surfaces a transition touches. In this way, our model provides an explanation for the effectiveness of prayer.

On a more familiar scale, let us consider the relation between a person and a campaign. At this level, the relationship is more similar to that of a body to an organ system than a cell.

In engaging with campaigns there are both objectifying and subjectifying attitudes that can be taken. An obvious strategy would be to maximally objectify the desired outcome or milestones, while maximally subjectifying the difficulties or obstacles that must be overcome.

Thinking about success in an objectifying way forces specificity, which is widely considered useful to achieving goals from a practical perspective. But holistic temporality predicts that a potentially more beneficial activity would be the extension and intensification of transition energy across a large time frame. This transition energy may be accessible for use in affecting the probable events at the borders of the campaign. Of course, an entanglement across such a long time stretch would make information available that could alter the decisions made at the outset. Often in conceptualizing the role of willpower in affecting outcomes we think of willpower used at the outset of a campaign. But holistic temporality models willpower as a four-dimensional quantity. The degree to which it would be effective at all would depend on the degree to which it was a single four dimensional holon, with information exchanged across the entirety of its four dimensional extant. Willpower used at the outset would effectively influence the outcome only to the degree that it is in coherence with the willpower extant through the rest of the campaign.

How one can practically bring the transition energy of a smaller temporal holon into coherence with that of a larger one touches on a fundamental question the holistic temporality model raises. To the degree that classical intuitions about time acknowledge entanglement between different moments at all, directly adjacent ones. But experiments such as the quantum eraser indicate that delayed measurements can have an effect on events not directly adjacent to the measurement itself. What then is the determining factor as to whether nonadjacent holons are sufficiently coherent so as to exchange information?

Research such as Sheldrake's may give an indication. If relative degrees of entanglement are based on the rate of information transfer, then although transfer happens more easily between temporally proximate regions it is not limited to them. Adjacent temporal regions are therefore more likely to exchange information and thus be in superposition from the point of view of an typical observer, but they are not the only ones capable of doing so. A human seeking maximum effectiveness within a campaign could therefore focus on information exchanged with the campaign at all its times of existence, even if they are separated by temporal regions unrelated enough to each be separate holons from the point of view of the other.

Information exchange sufficiently rich for entanglement to occur, with the benefits in probability extension this brings, is characterized by the presence of subjective experience. Although the precise neural mechanisms that give rise to subjective experience requires further study, maximizing subjective consciousness by directing attention to the subjective experience of all parts of a campaign should increase the total consciousness, and thus the transition energy available. Ensuring all this energy 'pulls in the same direction' is then a matter of the desires of all subregions of the campaign being aligned.

How to align these desires is first a question of knowing what they are. Deep familiarity being a prerequisite for group flow makes sense in this context. The desires of all participants must be 'brought to the table' in order to allow maximum alignment of diverse desires. This sort of openness is often taken to extremes in very successful teams. For example, Dieter Duhm, cofounder of Tamera, has credited a significant part of that intentional community's longevity to the members willingness to be unusually open with one another, even about their sexual desires.

A key question about how a 'unity of purpose' is achieved across a long time span can be approached with the previously raised points. Consider the example of a human at prayer. A common feature of prayer is to emphasize the desire of the deific figure, 'thy will, not my will, be done' being a commonly expressed sentiment. If we accept the larger holon as having its own larger, looser neural net and consciousness, then accessing this larger consciousness would presumably allow the human at prayer to gain insights generated by the larger party's grander neural net. Aligning with this larger consciousness's desires is presumably necessary to be successful in one's endeavors. But so long as one's individual desires are not in conflict with the larger consciousness, adding them in to the flow can lead to an intensification of the consciousness at work.

Under the holistic temporality model, both expansion and intensification of consciousness takes place during subjective observation. Prayer benefits the human by allowing access to a greater consciousness unbounded by local concerns. But the universal consciousness also benefits, having the locally intensified consciousness of the human align with its own.

It is in the skillful interweaving of the greater and lesser consciousnesses that the potential for better living can be found. The human holon must have enough individuality - fostered by objective interior observation - to have desires of its 'own'. Then, these desires must be aligned with the desires of the greater mind through intersubjective observation. In this way the individual can not only benefit from, but intensify the experiences of, the greater mind.

References

1: Flow State

[1] Wolchover, Natalie (December 1, 2016). "Quantum Gravity's Time Problem". Quanta Magazine.

[2] Koestler, Arthur (1967). "The Ghost in the Machine". Macmillan

[3] Sudarshan, E. C. G.; Misra, B. (1977). "The Zeno's paradox in quantum theory". Journal of Mathematical Physics.

[4] Feynman, Richard P.; Robert B. Leighton; Matthew Sands (1965). "The Feynman Lectures on Physics, Vol. 3". Addison-Wesley

[5] Yoon-Ho, Kim; Yu, R; Kulik, S.P.; Shish, Y.H.; Scully, Marlan (2000) "A Delayed Choice Quantum Eraser". Physical Review Letters

[6] Oeckl, Robert (January 2004). "The General Boundary Approach to Quantum Gravity". Proceedings of the first internation conference on physics, Amirkabir University, Tehran

[7] Penrose, Roger (1989). "Shadows of the Mind: A Search for the Missing Science of Consciousness". Oxford University Press.

[8] Tononi, Giulio (2004) "An Information Integration Theory of Consciousness". BMC Neuroscience

[9] Lambert, Chen, Et.Al. (2013) "Quntum Biology" Nature Physics

[10] Anderson, Mark (2007). "New Experiment Probes Weird Zone Between Quantum and Classical". Wired Magazine

[11] Einstein, Albert (author); Lawson, Robert (translator) (1920). "Relativity: The Special and the General Theory" University of Sheffield

[12] McFadden, John; Al-Khalili, Jim (June 1999). "A quantum mechanical model of adaptive mutation". Biosystems

[13] Zurek, W.H. (1991). "Decoherence and the transition from quantum to classical" Phys. Today

[14] Frauchiger, Daniela; Renner, Renato (2018). "Quantum theory cannot consistently describe the use of itself". Nature Communications

[15] Zielinski et al (2005). "Quantum States of Atoms and Molecules" Journal of Chemical Education

[16] O'Neill Shermer L, Rose KC, Hoffman A. (2011). "Perceptions and credibility: Understanding the nuances of eyewitness testimony." Journal of Contemporary Criminal Justice.

[17] Mikhailovich Bitsilli, Petr. (1983) "Chekhov's art, a stylistic analysis". Arids

[18] Mischel, Walter; Ebbesen, Ebbe B. (October 1970). "Attention in delay of gratification". Journal of Personality and Social Psychology.

[19] Dietrich, Arne (December 2004). "Neurocognitive mechanisms underlying the experience of flow". Consciousness and cognition vol 13, issue 4.

[20] Gretzky, Wayne (apocryphal)

[21] StarCraft 2 Design Team (July 2020) "Reflections on a

decade: The best games of competitive Starcraft 2, Part 1: Wings of Liberty" news.blizzard.com

[22] Born, Max (December 1954) "The statistical interpretation of quantum mechanics", Nobel Prize Lecture

[23] Kotler, Steven (March 2014), "The Rise of Superman: Decoding the Science of Ultimate Human Performance", New Harvest

[24] Sheldrake, Rupert (September 2009), "Morphic Resonance: The Nature of Formative Causation", Park Street Press

[25] Cabello, Adan (2017), "Interpretations of Quantum Theory: A Map of Madness", Cambridge University Press

[26] Spottiswoode & May, "Skin Conductance Prestimulus Response: Analyses, Artifacts and a Pilot Study" Laboratories for Fundamental Research

[27] Honorton, Ferrari, and Hansen, "Meta-analysis of Forced-Choice Precognition Experiments (1935-1987)", Prepared for the US Army Medica Research and Development Command

[28] Olival Freire Jr., (2005) "Science and exile: David Bohm, the hot times of the Cold War, and his struggle for a new interpretation of quantum mechanics", Historical Studies on the Physical and Biological Sciences, Volume 36, Number 1

[29] Paz, Habib, and Zurek, (January 1993), "Reduction of the wave packet: Preferred observable and decoherence time scale", Physical Review D, Volume 47

[30] Benson, Beary, & Carol (1974) "The Relaxation Response"

Psychiatry: Interpersonal and Biological Processes, Volume37

[31] Hadley, Mark J, (1996) "The Logic of Quantum Mechanics Derived from Classical General Relativity" Foundations of Physics Letters 10(1) 43-60

2: Agency

[1] Baumeister, R.J., Bratlavsky, E., Muravan, M (1998) "Ego Depletion: Is the Active Self a Limited Resource?" Journal of Personality and Social Psychology 74(5) 1252-1265

[2] Lange, F, and Eggert, F, (2014) "Sweet delusion: Glucose drinks fail to counteract ego depletion" Appetite Volume 75, pages 54-63

[3] Faye, J. (2019) "Copenhagen Interpretation of Quantum Mechanics" Stanford Encyclopedia of Philosophy

[4] Timpson, C. (2008) "Quantum Bayesianism: A Study" Studies in history and philosophy of science part B: Studies in History and Philosophy of Modern Physics

[5] Kim, Y. Et Al (2021) "Quantum Biology: An Update and Perspective" Quantum Reports 3, 1-48

[6] Radin, D & Nelson, R (1989) "Evidence for consciousness-related anomalies in random physical systems" Foundations of physics 19, 1499-1514

[7] Radin, D. Patterson, R. and Mason, L. (2007) "Exploratory Study: The Random Number Generator and Group Meditation"

Journal of Scientific Exploration Vol 21, No 2, p295-317

[8] Mischel, W, (2015) "The marshmallow test: Why self-control is the engine of success" Academia.edu

[9] Radin, D.I., Nelson, R.D. (1989) "Evidence for consciousness-related anomalies in random physical systems" Found Phys 19, 1499–1514.

[10] Radin, D., Michel, L., Galdamez, K., Wendland, P., Rickenbach, R. and Delorme, A., (2012). "Consciousness and the double-slit interference pattern: Six experiments" Physics Essays, 25(2), p.157.

[11]1.Mohapel P. "The neurobiology of focus and distraction: The case for incorporating mindfulness into leadership". Healthcare Management Forum. 2018;31(3):87-91. doi:10.1177/0840470417746414

[12] Penrose, Roger (1989). "Shadows of the Mind: A Search for the Missing Science of Consciousness". Oxford University Press.

[13] Cabello, Adan (2017), "Interpretations of Quantum Theory: A Map of Madness", Cambridge University Press

[14] Hartman, Ben (2015), "The Lean Farm: How to Minimize Waste, Increase Efficiency, and Maximize Value and Profits with Less Work" Chelsea Green Publishing

[15] Kotler, Steven (2015), "The Rise of Superman: Decoding the Science of Ultimate Human Performance", Quercus Publishing

[16] Oakley, Barbera (2018), "Learning How to Learn: How to Succeed in School Without Spending All Your Time Studying; A Guide for Kids and Teens", TarcherPerigee

[17] Rospars, Jean-Pierre and Meyer-Vernet, Nicole (2021) "How fast do mobile organisms respond to stimuli? Response times from bacteria to elephants and whales" Physical Biology, Volume 18, Number 2

[18] Baumeister, Roy F. and Tierney, John (2012) "Willpower: Rediscovering the Greatest Human Strength" Penguin Books

[19] Josipovic, Zoran, (2013) "Neural correlates of nondual awareness in meditation" Annals of the New York Academy of Sciences

3: Prayer

[1] This was revealed to me in a dream. Congratulations on making it this far, if this study provoked any original thoughts I hope you will share them with myself or others.